WOOD

for
Joyce and Wallace Gibson
Régine Chopinot
and
Julia

WOOD

Andy Goldsworthy

Introduction by

Terry Friedman

VIKING

VIKING

Published by the Penguin Group
Penguin Books Ltd, 27 Wrights Lane, London W8 5TZ, England
Penguin Books USA Inc., 375 Hudson Street, New York, NY 10014, USA
Penguin Books Australia Ltd, Ringwood, Victoria, Australia
Penguin Books Canada Ltd, 10 Alcorn Avenue, Toronto, Ontario, Canada M4V 3B2
Penguin Books (NZ) Ltd, 182-190 Wairau Road, Auckland 10, New Zealand

Penguin Books Ltd, Registered Offices: Harmondsworth, Middlesex, England

First published in Great Britain by Viking 1996

10 9 8 7 6 5 4

Produced by Jill Hollis and Ian Cameron
for Cameron Books,
PO Box 1, Moffat, Dumfriesshire DG10 9SU, Scotland

Filmset in ITC Stone Sans Medium by Cameron Books, Moffat
Colour reproduction by Alfacolor, Verona
Chlorine-free paper made by Leikam, Austria
Printed in Italy by Artegrafica, Verona

A CIP record for this book is available from the British Library

ISBN 0-670-87137-0

Works not captioned in main text

p.1

Maple leaf throw

KOSHIN-GAWA VALLEY, ASHIO, JAPAN

3 NOVEMBER 1990

p.2

Stacked sticks

Assisted by Cécile Panzieri and Peter Raczeck

CENTRAL PARK, NEW YORK

JUNE 1993

p.22

Woven
windfallen
redwood branches

FAIRFAX, CALIFORNIA

JUNE 1995

p.36

Sand
brought to an edge
to catch the morning light

MOUNT VICTOR STATION, SOUTH AUSTRALIA

17-18 JULY 1991

p.48

Raining heavily
long grass from the field edge
wrapped branch

RUNNYMEDE, CALIFORNIA

28-29 OCTOBER 1992

p.120

Roadside poppy petals
held with water
to horsechestnut leaf
late evening
calm

DRUMLANRIG, DUMFRIESSHIRE

6 AUGUST 1992

'Two Autumns' exhibition organised by Tochigi Prefectural Museum of Fine Arts and Setagaya Art Museum, Japan, curated by Eriko Osaka, Hiroya Sigimura and Yukihiko Ishii – endpapers; leaf throw p.1; grass throws p.28; grass screen pp.28-29; horsechestnut screen p.80; installation pp.82, 83.

Central Park project organised by Galerie Lelong, New York – stacked sticks p.2.

John and Dodie Rosekrans, Runnymede Sculpture Farm, Woodside, California – dust throws, p.17; clay holes pp.20, 21; installation San Jose Museum of Art pp.42, 43 – clay holes commissioned by John and Dodie Rosekrans at Runnymede; branch and hole pp.52-53.

'Mid Winter Muster', project commissioned by Adelaide Festival and Adelaide Botanic Gardens, Australia – sand holes pp.18-19; sand work p.36; mulga branches pp.56-57.

Project at Digne-les-Bains, France, organised by Nadine Gomez-Passamar and Guy Martini, a collaboration between the Musée de Digne and the Réserve Naturelle Géologique de Haute Provence – cairns pp.7-13; stone house p.30.

'Breath of Earth' exhibition organised by San Jose Museum of Art, curated by Peter Gordon, assisted by Richard Karson – burnt sticks p.39.

'time machine' – two installations, part of a group exhibition at the British Museum, London, and the Museo Egizio, Turin, curated by James Putnam; London work sponsored by Roxie Walker, Turin exhibition sponsored by Iveco – pp.44, 45; leaf horn p.70.

Project organised by Laumeier Sculpture Park, St Louis, Missouri – hickory leaves, p.71

Wood through wall, commissioned by Joel and Sherry Mallin, pp.54-55.

Alaska project organised by the Alaska Design Forum and the Anchorage Museum of History and Art, curated by Peter Lipson and Dave Nicholls – stick lines and ice houses, pp.60-65.

California project organised by Haines Gallery, San Francisco – screen of rushes, p.81.

'Végétal', a production by Ballet Atlantique-Régine Chopinot, premiered in November 1995 in association with La Coursive, Scène Nationale La Rochelle, Théâtre de la Ville, Paris, & Festival Sigma, Bordeaux. Choreography by Régine Chopinot, scenography by Andy Goldsworthy (supported by the Ministère de la Culture, Délégation aux Arts Plastiques et Direction Régionale des Affaires Culturelles Poitou-Charentes), soundtrack by Knud Viktor, lighting by Maryse Gautier; dancers: Rebecca Adam, Dimitri Chamblas, Régine Chopinot, Guillaume Cuvilliez, Marie-Françoise Garcia, Virginie Garcia, Gilles Imbert, Hiroko Kamimura, Samuel Letellier, Georgette Louison Kala-Lobé, Élodie Pallaro, Esteban Peña Villagran, Jie Peng, Michèle Prélonge, Duke Wilburn – pp.14, 16, 32-33, 41, 58, 59.

Cameron Books is grateful to Common Ground (Seven Dials Warehouse, 44 Earlham Street, London WC2H 9LA) for permission to reprint the text entitled 'Leaf' in this book from *Leaves* (1989).

Jacket photograph by Judith Goldsworthy.

Andy Goldsworthy is Senior Lecturer/Practitioner-Fine Art for 1996 at the University of Hertfordshire

The artist is represented by: Haines Gallery, San Francisco; Michael Hue-Williams Fine Art, London: Galerie Lelong, New York and Paris; Galerie S65, Aalst, Belgium.

Capenoch Tree series made with the goodwill of Robert and Margaret Gladstone, pp.84-119.

Thanks to the Buccleuch Estates for their continuing support and to Elinor Hall for her administrative help.

I am especially grateful to Wallace and Joyce Gibson for their help in making works around the Capenoch Tree in all weathers and at all hours.

Opposite page:

Slate stack

STONEWOOD, DUMFRIESSHIRE

1988

CONTENTS

STONEWOOD

Terry Friedman

Andy Goldsworthy makes sculpture in the landscape using the materials of nature immediately to hand and the chance conditions of place, time, weather, season. The now-familiar forms of his art – arches, circles, columns, domes, holes, lines, spheres, spirals, spires – are powerful expressions of the patterns and rhythms of growth. They are attempts to understand the purpose of sculpture and through it the purposes of nature itself.

Stonewood is a small strip of wild woodland lying along a stretch of steeply sloping river bank near Penpont, in the Scottish county of Dumfriesshire, where Goldsworthy has made his home since 1986. In 1988 he built here a high, nearly square wall of horizontally stacked slates with a circle laid vertically inside, as a protective sentinel to an oak tree which had been burnt at its base (p.5). In a sculpture made in November 1993 in New York (pp.54-55), two long sections of a fallen tree trunk have been embedded horizontally in an old dry-stone wall, reworked and divided to form a gate in a wood; the tension between the compressing property of stone and the expanding nature of tree as revealed by the annual growth rings on the exposed ends of its limbs is palpable.

This dialogue between stone and wood remains an abiding concern of the sculptor. In recent years, working increasingly on and around trees, he has come to recognise profound, elemental differences between the two materials: 'A stone is passive, a witness to the place in which it sits; it is a focus, the core, the remains of something that was larger; its movement is one of erosion . . . A tree is an active part of its place, it makes that place richer and is an indication of the way something can change a place – in this respect it is a lesson to me about my own life as a sculptor. A tree is close to the gesture of a leaf or stone throw, but over a much longer span of time, during which it thrusts up and reaches out into space before decaying and dropping back to the ground'. The analogy here is between nature (tree) and man (the sculptor): roots/feet, trunk/body, branches/arms and hands, leaves/the throw.

Yet stone and wood also interact, possessing interchangeable properties and an unsuspected but real reliance on each other. Sequential arrangements of stones of different colours give a feeling of something flowing through the stone, like sap through a tree. Works constructed of stacked stones or stacked slate express 'the fullness, vigour, heavy ripeness and power of nature generated from a centre deep inside' and have an energy like that of tree growth. Working at Digne-les-Bains, in the remote Basses-Alpes of Provence, during the summer of 1995, Goldsworthy became aware that 'just as the mountains hold the stone, the trees hold the stones of a mountain. It appears that the stone is the stronger but somehow it is protected by the tree. I like the inversion of the idea that stone is hard and strong, and the tree is organic and fragile'. A stone column enclosed by branches (p.30), made on the pebbled bed of the River Bès, with the woods behind and behind them the mountains, was completed as the sun went down and the light became softer and the shadows stronger: 'the shadow cast by the bright sun is too harsh. I wanted something more intense as if generated from within'.

Goldsworthy's usual practice has been to make each sculpture, which can vary in size from miniature to monumental, using only one kind of material and constructed during a brief time span, often within a single day. The work is then photographed before it naturally disintegrates. Recently, however, there has been a new departure. On the Bès at Digne he had the idea of making a solitary cairn (a traditional, man-made pile of stones of ancient Celtic origin that served as a landmark or boundary-marker) that grows and changes day by day, layer on layer: a core of yellow stones, then grey-blue stones, stones changing from reddish-brown to orange to yellow to pale yellow to white, yellow stones and blue-grey stones together, bleached, burnt-ended stalks, graded stones which are then wet, grey stones and sticks, large, concentrically arranged stones, a cocoon of sticks. 'I have always felt that each work is made on the last, but never actually made one in a way that the piece does not

just replace the last but grows from it'. He enjoyed the economy achieved by this approach. 'In the past when I have constructed piles there has been so much effort in making the infill for the next work.'

As the cairn, growing over nine consecutive days (17-25 July 1995), reached its full height of around seven feet, its relationship to the distant Mont St Vincent became acute for Goldsworthy, and the two complimentary structures seemed cyclical inversions of each other. The layered growth of the cairn, with its core of yellow stones dating from the earliest geological strata of the mountain (about 250 million years old); the bleached and burnt stalks laid on a slope to echo angled strata of the mountain; the very shape of the completed cairn itself suggesting that it is 'the tip of something bigger below ground', a miniature reflection of the mighty mountain in the exultation of its youth. The gradual wearing down of the mountain over millenia by wind and water, ultimately to expose its bottom-most aureate strata : the infinitely swifter and inevitable disintegration of the cairn by a rising river slicing through its circular layers, exposing its growth rings, like the rings of a tree.

These interactions between the processes of growth and decay, construction and deconstruction, are at the heart of Goldsworthy's installations for the ballet, 'Végétal', commissioned in 1994 and premiered in November 1995 by Ballet Atlantique-Régine Chopinot at La Rochelle in western France. The continuously evolving focus of the dance is embodied in the responses of fourteen dancers to sculptures built, dismantled and re-formed by them on stage, accompanied by Knud Viktor's soundtrack of taped natural sounds. The installations at the Setagaya Art Museum, Japan (19 February – 27 March 1994) and at the San Jose Museum, California ('Breath of Earth', 5 February – 23 April 1995) had been presented by the sculptor as a series of related sculptures, layered one behind another. So a collaboration with contemporary dance probably came as no surprise to him, nor should it be for us. For Goldsworthy 'dance is the performing art that is closest to sculpture. Its use of the human body, the essence of movement. I believe that there is a dance that happens betwen sculptor and material that I wanted to be expressed in 'Végétal'; it's also a dance in poetry that I see in many other manual activities – dry stone walling, hedge laying – the rhythm of making things. I feel affinity towards that as a way of working'. Over the years some of his sculptures have involved balletic body actions: rainbow splashes, stick throws, stone throws, leaf throws, dust throws. 'That is the work that I have always considered to be the essence of what I do – the body as the sculpture. I've always seen myself as an object in the work; that I'm nature too. With the throws the human element is central, so the idea of dance – human energy, human nature – is something that I can respond to very, very strongly'. From time to time, while working out of doors, Goldsworthy lies spreadeagled on the ground at the outbreak of rain to leave a 'shadow' of his body on the wet earth. At Digne, this became a near-visionary experience when it rained heavily, each drop making 'a splash on the stones two or three inches across. The sun was shining and they appeared like metal filings dropping from the sky'.

Goldsworthy came to the dance collaboration with strong feelings. 'I didn't want the production to be too pastoral, to have a back-to-nature, New Age feel. It should not be ritualistic. I didn't want my contribution to be merely a backdrop or "prop" standing inert on the stage. I felt that a collaboration should be a fuller one, a stage that does not sit there but which generates the dancers' movements, which is made by the dancers, that makes the dancers'. Régine Chopinot made the dancers perform as if they were not moving themselves but being moved by some other force. In 'Végétal', which lasts approximately one hour and forty minutes, sculptor and choreographer wanted the materials with which the dancers dance – stones, branches, earth, leaves – to be the actual materials of sculpture. On one of his preparatory visits to La Rochelle, Goldsworthy took the entire company, including many of the administrative staff, to a wood. There they collected sticks, built a dome and then

dismantled it to form a ring. Later, on a visit to Penpont in 1995, they threw sticks and earth (p.59). The audience was also to experience the actual making of sculpture as an essential component of the dance, the making to take place on stage in 'real' and unpredictable rather than 'theatrical' time. 'Real' time would be made manifest to the audience symbolically, but unambiguously, by a lone dancer (Régine Chopinot herself), who from the start of the ballet to its finish would with measured slowness circumnavigate the rectangular perimeter of the stage clockwise, like the minute hand on a watch face, as if stating 'this is my path, my journey, my time'. Unpredictable mishaps during the construction of fragile and vulnerable sculptures on stage by some of the dancers, who are personifications of the sculptor himself, would have to be accommodated as part of the performance. The making of sculpture, which for Goldsworthy had always been a private activity, was now made public.

Végétal, the French word for plant (as distinct from animal and mineral), but here given the additional vital meaning of growth, is a dance about beginnings, unfolding, breathing. The plant world provides the oxygen essential to life. The growth of the sculptures on stage is like the growth of a tree, starting with a thrust upwards and then a spreading outward, branch-like. The stone column built on stage is like the image of the stone column at Penpont (pp.110-111) 'anchored in the core of the tree, the heart of the tree where the energy is'.

The dance is divided into five continuous, interlinked episodes: Earth-Terre, Seed-Graine, Root-Racine, Branch-Branche, Leaf-Feuille. Since this is also the format adopted in the present book, *Wood*, which has the additional section, Tree, at the end of the cycle, we can explore the dance in the wider context of Goldsworthy's recent work elsewhere, including sequential sculptures, such as the branch line at Anchorage, Alaska (25 November – 9 December 1995, pp.60-63) and the Capenoch tree (March 1994 – February 1996, pp.84-119).

Earth

'Végétal' begins with Earth, in dark and dusk; the sounds, constructed on the theme of a skylark's song, are amplified, warm and heavy. A soft, faint, white light falls on the stage floor, surrounding an ambiguous inner circle of black, perhaps some inexplicable object, or a deep hole generating energy. Holes dug in the earth have preoccupied Goldsworthy since the 1970s: 'The black of a hole is like the flame of a fire. The flame makes the energy of fire visible. The black is the earth's flame – its energy'. At Runnymede (California), near a path in the wood, in the newly moist and soft clay following years of drought, he excavated a trail of holes (p.21). These were made in conjunction with a nearby tree stump embedded with a series of elliptical holes in impacted clay, a sensual object redolent of fecundity (p.20). Nourished by a sudden shower, young, bright-green blades of grass emerged from the black holes.

In 'Végétal' the floor is spread with a thin layer of red earth. For Goldsworthy the iron content present in blood, which allows us to breathe, and also in stone, is manifest as red, 'red flowing around the Earth as a vein'. Three dancers lying in a triangle round the 'hole' gradually awaken, breathe, curl over, unfold, warming their bodies. They rise and throw handfuls of red earth which float above the stage 'like a spirit hanging about' (p.14).

Seed

A lone dancer, centre stage, builds a life-size column of stones (these are the mysterious contents of the inner circle): carefully graded and stacked, precariously balanced (pp.32-33). It is a construction of vulnerability not virtuosity; you are aware of the tension in the audience. (At the Paris performance attended by the writer the upper tiers of stones fell during construction, crashing to the floor, surprising the audience by their actuality, and had

to be patiently rebuilt. This has happened many times before to the sculptor working in the landscape, his way of working needing randomness, unpredictability, even temporary loss of control. These qualities are understood by the choreographer.) When the column stands complete Goldsworthy finds the counterpoint between column and dancer 'just beautiful', associating this and the icicles in the stick domes made in Alaska (pp.64-65), with Matisse's *Backs* (a series of four large bronze reliefs of nude female figures made over several decades), where the vertical flow of hair becomes the spine. Now six dancers, bellies and rumps tattooed by the red marks of the earth, gather loose stones, weigh them, pass them back and forth from right hand to left hand, crisscrossing the stage in a 'germination dance'. They interact with each other and with the balanced column, which is the equivalent of the human spine and which sustains their energy. Goldsworthy, too, has discovered an energy in stone that is like 'a seed that becomes taut as it ripens, often needing only the slightest of touches to make it explode and scatter its parts'. He has used the stone/seed analogy in other sculptures, and in drawings where snowballs, stained with dye made from boiled ash tree seeds, are placed on large sheets of heavy paper, melt and grow like the roots and branches of a tree (p.25).

A dancer then breaks forward, topples the column and in swift, annular movements trips the other dancers. Sounds are the sounds of stones falling, hitting each other; the quiet voice of erosion: 'very fine particles, between splinters of stone and sparks of fire'. Pairs of dancers fall and rise and fall and rise, in what Chopinot calls 'punching duets'. Even after the column has collapsed, they 'keep its gravitational force inside them. Life and seed are in the belly'.

Root

The back wall of the stage is illuminated to reveal a huge white canvas filled with a complicated, continuous, serpentine drawing made of dried ferns stuck with rabbit skin glue. Here Goldsworthy's idea of layering is of revealing something lying beneath the surface, a memory within the wall, a memory of the drawing's root origin. In an impressive photograph of a work made at Scaur Glen, Dumfriesshire (p.40), he reveals the compelling structural oneness of root (snaking bracken fronds laid on the forest floor). Consistent with his sequential way of working, outdoor earth, sand and leaf serpentines are brought indoors and constructed as installations. For *time machine Ancient Egypt and Contemporary Art*, organised by The British Museum's Department of Egyptian Antiquities in 1994, he built a long and gradually rising mound of damp, compacted sand which snaked its way through the main sculpture gallery: 'Sand is somewhere between stone and earth. It can be compressed hard and yet it can be fluid. It has a sense of strength, fragility and movement. The work [flows] through the room – touching the [ancient] sculptures and incorporating them into its form to give a feeling of the underlying geological and cultural energies that flows through the sculptures' (p.44). The work was made for a day, photographed and removed, yet retained in the public exhibition as a memory, 'ghosted' by two small, serpentine sculptures of sand and sweet chestnut leaves, each laid in the hollow of an ancient stone sarcophagus displayed in the same gallery. 'A work made with leaves is a celebration of growth, yet cannot work without expressing some anticipation of death, in a way that understands that death is a part of growth. The sarcophagi are not just containers of death, they are containers of life, in that out of death comes life'. A daring variation of these resurrectional images was made in the following year for a version of *time machine* held in the Museo Egizio at Turin in Italy (p.45). Here the temporary sand sculpture, photographed and removed, as before, was shadowed by an equally monumental drawing of bracken fronds pinned with thorns to the vaulted ceiling directly above.

The moment of photography at Turin was very specific. Goldsworthy found that the temporary sculpture was 'brought to life for about an hour around midday as the sun moved down from one end of the gallery to the other, the edge of sand catching and working the light. Yet I know that after I left, when the sculpture was on view before being dismantled prior to the exhibition, the artificial light would have killed off that quality and form, and no-one would see it in its proper light. I feel it is inappropriate to give people precise instructions on when to see a work. Instead I take a photograph which becomes a personal explanation of how and when the work should be seen. This is my view of my work – it does not exclude other views or the possibility that the work will go on to generate other responses, but for that moment this for me is the work.' For Goldsworthy the photograph plays as vital a role in presenting indoor sculptures as it does ephemeral outdoor work, and he has discovered a compelling relationship between the two. During the particularly cold weather experienced in south-west Scotland in February 1996, he constructed a spiral of broken icicles, frozen with icy water, which wound itself round the trunk of a tree (p.67). 'For some time it appeared as if the work was in the only place illuminated by the sun; it glowed with such an intensity that it was as if the icicle was both absorbing and generating light. This was the moment when the work came alive. This is what I feel Brancusi must have felt when he photographed the polished bronze bird as the light caught it in his studio. This intensity can only be shown for a short time. In fact, the moment is intense only because it lasts for a short time, and it would be wrong for such an intensity to last longer than that – it is latent in the work, otherwise it would become too much.'

Branch

Ten dancers process in circles, offering branches of gradually diminishing length and circumference to a male and female standing centre stage (at the spot previously occupied by the stone column), who construct around themselves a seven-foot high dome (p.58). A tree, too, builds itself layer by layer, its rings containing and protecting 'the ebb and flow of energy within'. Goldsworthy has encased the trunks of still-living trees in sticks and in stones (pp.46-47), and orbited them with stalk lines and ice swirls (pp.66-67). He has enveloped a branch still attached to a tree with poppy petals, a red vein breathing through the green of growth (pp.50-51). In Alaska (26-29 November 1995), he broke and froze together a line of branches, which was then incorporated into other lines, one work embedded in another, finally evolving into an ambitiously long zig-zag of branches strutting across the freezing snow and ice, taking on the quality of a drawing (pp.60-63). The following day and again a week later, he built domes of branches (a memory of 'Végétal'). In both, water was dripped from the top inwards to form an icicle two or three feet long (pp.64-65). 'The icicle is like the water contained within the tree; the tree becomes a house that protects the sap, the life blood of the tree. The icicle is like the spine of a dancer, connected to the spine of stone built at the centre of "Végétal". ' From the sculptor's Alaska diary: 'I never realised how interesting it is to watch an icicle form and how sensitive the line an icicle takes is to the weather, the wind. If it was breezy the icicle would veer off. I wanted a straight icicle, so I had to protect the tip from the wind, pouring so that it fell straight down.'

During the construction of the dome in 'Végétal', followed immediately by its deconstruction, which takes place in 'real' time and lasts altogether about twenty minutes, sculpture seems to dominate dance. From the gradually diminishing layers re-emerge the pair of incarcerated dancers, totem-like and rotating: they are the sap of the tree, the spine which had been the column. There is a feeling of gravity perfectly poised and balanced, like the Alaskan icicle within its dome of frozen branches. Sound is composed of 'fifteen chosen drops of water, and the appearance

of a fly from time to time; running water, hardly audible, as if the source of the drops was this sound of water'. The dancers unfurl the branches to form a ring, which is then re-formed into a ring of greater circumference. Within this the dance continues.

Leaf

Leaves are for Goldsworthy 'generated from the tree's core. They are the tree's senses, an expression of its vigour'. This final episode is described by the choreographer as a sculpture with the dancers' energy as the material, their bodies swirling in circles, turning until they fall. 'The tree is bare, loses its foliage, turns back within itself. Full cycle'. Large bundles of dried leaves are brought to the centre of the stage and deposited in a cairn-like mound. The full company dances round the mound in palpitating actions to the rhythm of breathing, like a pair of lungs; to 'the sound of the leaves', 'the steps of the dancers on the leaves', the crickets 'with their nocturnal ways'. The dancers do a lot of leaf throws. The stage suddenly returns to blackness. The ballet ends.

Goldsworthy's previous leaf works were made on the floors of woods, or laid in streams and on trees, and this continues to be a productive practice. But recent works using leaves, like some other of Goldsworthy's sculptures, have explored further compelling relationships with the land. The ambiguity he found so pleasing in presenting the serpentine fern backdrop on a vertical plane in the Root episode of 'Végétal' was because it represented both the leaf and the root, 'a reminder of the plant above and below the ground'. A sweet chestnut leaf spiral, in earlier sculptures harbouring cornucopian associations, is now placed at the base of the tree to signify root (p.68). Another leaf spiral cradled at the fork of trunk and branch and contiguous to the growth rings revealed on the stump end of a limb, makes an unexpected, powerful link between root and leaf as the foundation and summit of the process of growth. Single elm, maple and horse chestnut leaves still attached to their trees are wrapped delicately in poppy petals as if the blood of the sap has circulated through the stem into the tips of the fingers (p.120).

Tree

A magnificent oak tree at Capenoch, near his home in Dumfriesshire, which Goldsworthy has so far worked through two cycles of seasons (1994-1996), allowing him to concentrate on a sequence of works over an unusually extended period rare in his often frenetic life as a sculptor, has produced a new body of varied but closely related sculptures (pp.84-119). The tree is remarkable for having birthed an exceptionally precocious horizontal branch which hovers close to the ground, a leviathan surviving only by being welded at one point to another branch crossing overhead. A branch taking a chance, in turn propelling the sculptor towards unimagined but welcomed risks: the upper ridge of the branch leapfrogged by arches of stone, by a regiment of snowballs. In Runnymede wood, he had cut a long, narrow, sharp-edged and dark trench in the earth to 'shadow' the line of a fallen and isolated branch (pp.52-53). But the intimacy between branch and earth discovered at Capenoch taught him that 'the tree is the land, that the branch is like landscape, just as the tree itself is landscape'. So he began working the pregnant interval of space between them, the sculptures often 'untouched by the branch but very much about it'. He built a Digne-like snow cairn, first covering it with black mud, then half snow and half mud, and finally with untrodden snow. Aware, too, that he often works best when a tree is leafless and reduced to its structure, he built a stick cairn up and over the naked branch, with a circular opening exposing the bisected limb, as if looking into the body's skeleton. This is a sculpture similar to the Digne stone column enclosed by sticks (p.30) and the Mallorcan stone-cairn-enveloped olive trees (pp.46-47).

The core of the Capenoch cairn was a previous ice work which had, in the nature of things, collapsed. Constructed of stacked slabs of ice retrieved from a nearby pond, this sculpture (pp.90-91) took the form of a pointed arch, which rose from the ground, cut through the branch and peaked atop it. The branch faces east and is lit by the morning sun: dazzling light illuminated the arch like some miraculous image. Goldsworthy was unaware of a long-standing architectural theory that ascribes the origins of Gothic cathedrals to naturally growing avenues of trees: its columns trunks, its capitals leaves, its vaults branches ascending and crossing over the processional way. In 1772 Goethe eulogised Strasbourg cathedral, 'It rises like a most sublime wide-arching Tree of God, which with a thousand boughs, a million twigs, tells forth to the neighbourhood the glory of God'. Such evocative parallels begin to reveal Goldsworthy's place in a history of artists responding to the structures and mysteries of nature.

Cairn
stones, sticks, fire, earth, air, water
new layers added each day
materials gathered nearby
collecting became harder as the pile grew larger
hot

DIGNE-LES-BAINS, FRANCE
JULY 1995

Works discussed in the text and illustrated in this book may be found on the pages noted in parentheses after references to them; for those not illustrated, *see* T. Friedman and A. Goldsworthy (eds.), *Hand to Earth: Andy Goldsworthy Sculpture 1976-1990*, Leeds City Art Gallery, 1990/Abrams, 1993, pp.24-42 (holes), 43-51 (throws and splashes), 142-159 (outdoor monuments); A. Goldsworthy, *Andy Goldsworthy*, Viking/Abrams, 1990 (rain shadows in Holland and Scotland, 1984, Japan, 1987, and Cumbria, 1988); A. Goldsworthy, *Stone*, Viking/Abrams, 1994, pp.34-47 (cairns); *Black Stones Red Pools*, 1995 (red earth). Statements by the artist for the Egyptian work are extracted from the exhibition catalogues *time machine Ancient Egypt and Contemporary Art*, The British Museum, 1994, pp.5-6, 46-49 and *time machine Antico Egitto e Arte Contemporanea*, Museo Egizio, Turin, 1995, pp.24-27. The author has made use of the sculptor's unpublished diary notes for work at Runnymede, California (25 October – 4 November 1994), Anchorage, Alaska (25 November – 9 December 1995) and Capenoch, Penpont (January and 3-4 February 1996). The text for work at Digne-les-Bains, France (14-26 July 1995) is based on the sculptor's unpublished diary notes and a letter to the author dated 19 February 1996 from Nadine Gomez-Passamar of the Musée de Digne. The author attended the 21 February 1996 performance of 'Végétal' at the Théâtre de la Ville, Paris; his text is based on an interview with the sculptor conducted at Penpont on 12 November 1995, further discussions on 13-14 February 1996, and *Végétal* (January 1996), an illustrated programme for the Paris performances containing statements by Goldsworthy and designed by him and Patrick Barbanneau. For the Goethe reference, see G. Henderson, *Gothic*, 1967, p.181.

EARTH

'Végétal', 1995

I have worked with red earth in many places. It flows around the earth as a vein. It is red because of its iron content which is also why our blood is red. The red in blood is related to breath, and breath is what we share with vegetation.

In 1994, I made dust throws in California. I called these works 'Breath of Earth'. They had a quality of breath on a cold day; California is a place where you can feel the earth breathing – often violently.

Japan, 1993

I have become more aware of the significance of red and the intensity in the way it is revealed. I have experienced brilliant autumn colours in many places, but never a red so deeply disturbing as the Japanese maple. The tree is often isolated against a green mountain, appearing like an open wound. It is this contrast which gives the red its energy and power.

Mount Victor, South Australia, 1993

This place will always be important to me. It is difficult to define the light, the colours. I cannot describe the effect that red has on me. It is as deeply moving spiritually as red maples in Japan or the spring green grass in Britain . . . I have tried to touch that colour not just with my hands, but also with light. To understand that colour is to understand something of the spirit of this place.

California, 1992

I can't remember getting so wet, even in Scotland. It was a very heavy rain. Rain after a long drought is a powerful thing. The ground is very soft which is in such contrast to how it was when I first arrived. It was so hard. Now it is soft, deeply soft. Deer tracks dig into it deep. I made a work with holes, a trail of holes. The rain has sunk in so far and the earth is soaked thoroughly in a way that can only be because it has been so dry. I can see small green plants germinating and growing after the rain. There is a greenness coming through already, very fast. For me, there is something deeply interesting about a bright green grass blade growing out of a black hole. I have never experienced such vigorous growth in such a short space of time, not out of barren-looking earth. I now realise the profound impact of rain. That the soil reacted so vigorously to it is evidence of the energy that the rain released and the change that it initiated.

'Breath of Earth' – dust throws

Photographs Todd Hosfelt, Mary Maggini

RUNNYMEDE, CALIFORNIA
JUNE 1994

Ballet Atlantique

PENPONT, DUMFRIESSHIRE
AUGUST 1995

Hard sand
carved out with a stick

MOUNT VICTOR, SOUTH AUSTRALIA
15 JUNE 1991

Clay
worked into the hollow
of a tree
facing south
lizards

RUNNYMEDE, CALIFORNIA
3 NOVEMBER 1992

Holes dug in a soft wet earth
overcast
a few days later
grass coming through

RUNNYMEDE, CALIFORNIA
30 OCTOBER 1992

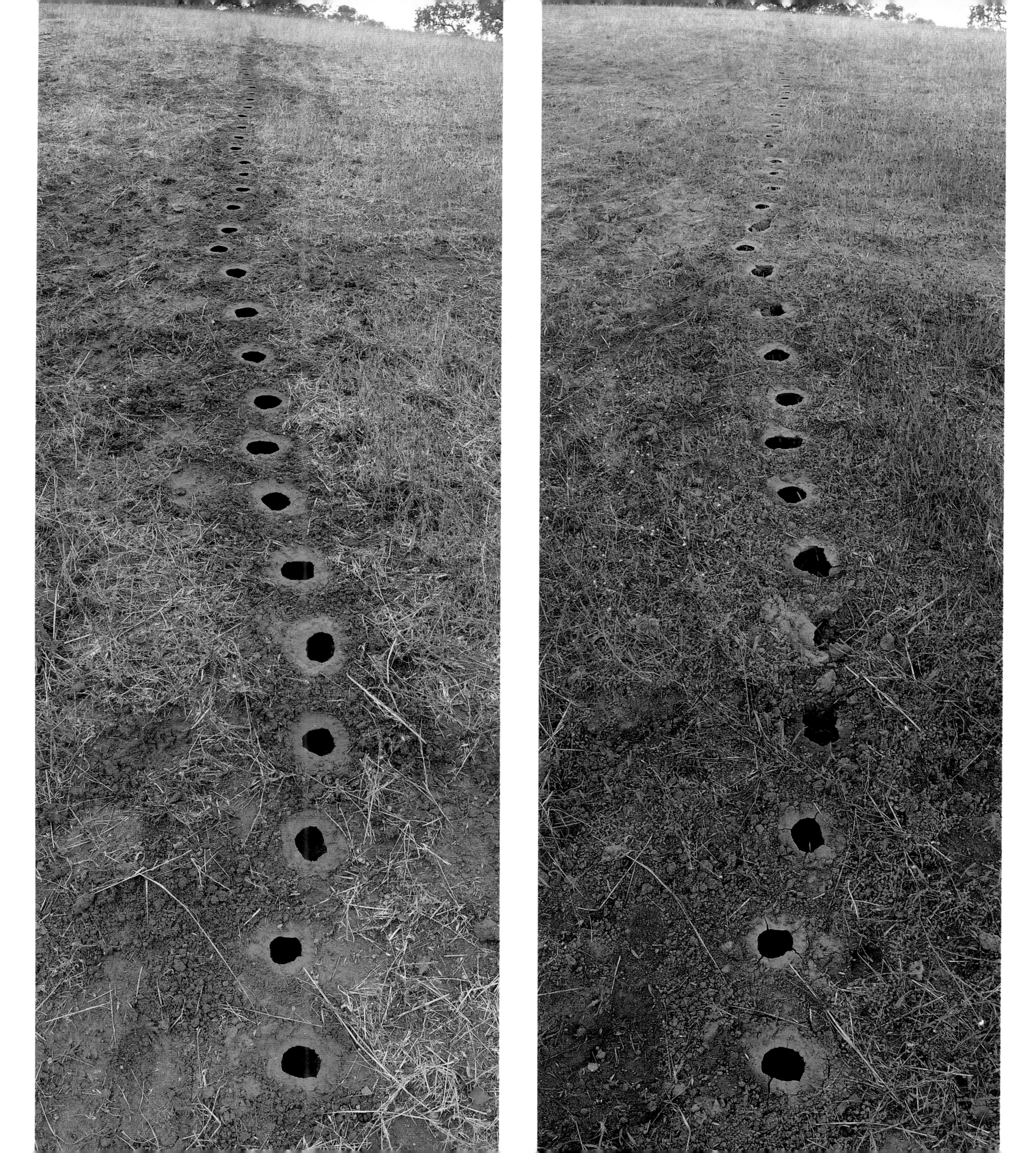

SEED

Mallorca, July 1994

These trees are so old they feel like stone. The stones have given growth to the olive trees; they grow out of the earth because of the stone – within the wood is stone. Cracked and weathered, with holes and openings; some with almost open trunks revealing their guts. You can see the internal tree. The stone column balanced as a partner to the tree. The stone grows within the tree – the seed. The column is a growth form, progressive. The tree is stone expressed in wood. One column made in the tree itself – in its hollow, heart. The tree becomes stone-filled. Although the light is a changing one, it is relatively predictable and I can wait until the following day to pursue the light that interests me for the work. Most of the work has been made in the early morning or late afternoon, in the soft light before the sun has risen or after it has gone down. Bright sunlight makes the form too much about surfaces. I need a quieter light to bring out the edge between stone and tree. This tree more than others has a quality of the stones having been drawn up through the roots as if they were fluid. This is why small stones are needed en masse to give the necessary movement, like a flock of birds or a shoal of fish. The individual is lost in the flow. After the random work of the past few years – throws and splashes – I can feel myself drawn to the more formal. This is necessary, I suppose, for if I pursue the random and chaotic my work would become stylised rather than provoked by the energy of chance. The tree has become the driving force. Growth has an intensely powerful structure. It may generate chance, but at its core it remains tight and concentrated.

'Végétal', January 1996

Stone has shown me many things about the structure of growth. I have found an energy in stone that can best be described as a seed that becomes taut as it ripens – often needing only the slightest of touches to make it explode and scatter its parts. The precariousness of a balanced column is like the fine edge between success and failure – the tension of growth. There are inevitably more failures than successes. If the column collapses before completion the fall is destructive. After completion, the fall has the quality of an energy breaking out.

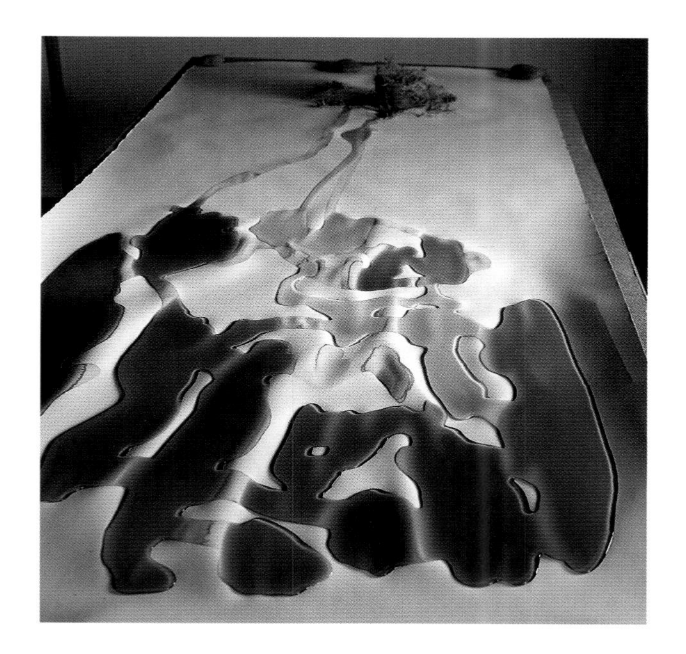

Snowball
gathered from below an ash tree
taken to my studio
stained with dye extracted from ash seeds
melting on paper

PENPONT, DUMFRIESSHIRE
FEBRUARY 1993

Beech tree snowball

SCAUR GLEN, DUMFRIESSHIRE

18 MARCH 1994

Oak tree snowball

SCAUR GLEN, DUMFRIESSHIRE

26 FEBRUARY 1994

Susuki throws

Photographs Ikuko Okada

OUCHIYAMA-MURA, JAPAN

21 NOVEMBER 1991

Susuki grass
held with thorns

TOCHIGI PREFECTURAL MUSEUM OF FINE ARTS, JAPAN

OCTOBER 1993

Stone houses

DIGNE-LES-BAINS, FRANCE

16 JUNE 1995

MT. KISCO, NEW YORK

MARCH 1995

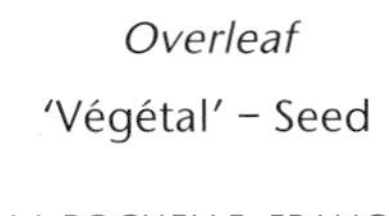

Overleaf

'Végétal' – Seed

LA ROCHELLE, FRANCE

NOVEMBER 1995

Balanced stones
olive trees

FORNALUIX, MALLORCA, SPAIN
JULY 1994

ROOT

'Végétal', January 1996

I am attracted to the line a snake makes as it draws the ground through which it moves – probing as it finds its way. These are qualities I also associate with root . . . I like the ambiguity of representing root on a vertical plane. It can be both the root and the leaf, a reminder of the plant above and below the ground.

Digne-les-Bains, France, 21 July 1995

What a day! Got up at 5.30 with my son Jamie. Just turning light. Went to the cairn. Made a fire from wood collected last night. Burnt the ends of carefully collected bleached branches. Made a cairn with a black burnt peak, very similar to the bracken root patch that I made two or three years ago: when pulling the bracken out of the ground I found it black. The fusion and exchange of energy between plant and earth had created the black.

Mount Victor, South Australia, July-August 1993

The work has travelled down the slope. It made connections to other lines: the pathways that I used to follow down hills, old shepherd tracks in the English Lake District which were sensitive to the line of the mountain and invariably the easiest way down, although not always the most direct. The edge followed the hillside down, catching the early morning light. Dead trees lying on the bank dictated the path of the work. On top of the bank stood a dead tree stretching into the sky – my lines of sand reaching into the earth below. I heard thumpings, like a heartbeat: I had worked on a warren. It must have been the rabbits inside stamping their feet.

Black-rooted bracken stalks
from the previous year
dead but still standing

SCAUR GLEN, DUMFRIESSHIRE
17 FEBRUARY 1990

Burnt sticks

SAN JOSE MUSEUM OF ART, CALIFORNIA
FEBRUARY 1995

Backdrop for 'Végétal'
bracken and fern fronds
held with rabbit skin glue

LA ROCHELLE, FRANCE
JULY 1995

Bracken fronds
stripped down one side
pinned to the ground with thorns
below an oak tree

SCAUR GLEN, DUMFRIESSHIRE
8 SEPTEMBER 1995

Runnymede clay
Scottish, Japanese, Californian bracken and ferns
held with thorns to the wall

SAN JOSE MUSEUM OF ART, CALIFORNIA
FEBRUARY 1995

Installations in London and Turin
each using local sand

EGYPTIAN SCULPTURE GALLERY, BRITISH MUSEUM, LONDON
22-25 OCTOBER 1994

MUSEO EGIZIO, TURIN, ITALY
20-27 NOVEMBER 1995

Proposals for tree and stone

1994

Olive tree cairn
late evening

FORNALUIX, MALLORCA, SPAIN
13 JULY 1994

BRANCH

Alaska, 28th November 1995

It is so slow. Each piece of branch has to be left to become firm before the next is added. For each that I gain there are several collapses. Gradually the work struggles upwards and yet I'm trying to achieve a line that flows effortlessly . . . Finally, as the sun was going down, I managed to rebuild the work back to the height it was at the beginning of the day. When the sun went down, I removed the supports and it held. It is extraordinary how the last, thin bit is so important in making the line alive. It is the tip that gives direction and energy to the line.

Alaska, 29th November 1995

The fine edge between success and failure is somehow expressed in this line of wood bound by the cold. There are 114 pieces in this line. Yesterday's work had 23 and the first had 12 sticks. Today's line is made in a space defined by many lines, the line of the inlet, the line of the mountains, the tree line. In making the line, I resisted using a single branch to achieve the curve. I wanted to make the curve out of smaller pieces rather than to find a curve already there. I want the line to be made up of wood, ice, wood, ice, wood, ice. Winter, summer, winter, summer, winter, summer. I like the idea of many pieces being joined together in a continuous line, just as the seasons are. Changes and the flow of time are lines that run through both tree and land.

All three works are rooted in the tree and in the place where the tree grows – they are growth lines. Today's line moves from the wood out onto the ice, and the other two from where the tree once grew. They explore the quality of wood. They bend and break. The curve and the angle. One harsh and the other flowing. All qualities in the material that is wood.

Alaska, 8th December 1995

Collected branches with help from several people. Made a stick dome. It was strange to be making this out on the ice when it contains so much of the memory of La Rochelle. In some ways this makes it stronger for me. At its centre, hanging from the roof, I have begun to make an icicle. I also made a variation on this work on the beaver pond in the wood. In each, an icicle will hang. This is so different to finding an icicle that I then freeze to a spot. These icicles grow in the place.

Alaska, 9th December 1995

I wanted an icicle that had the quality of a knife so that the wood enclosing it was almost like a sheath to the blade. Sharp, cold. These last two works are ones that I like very much. The branches conceal and protect the icicle. They are evocative of the fragile relationship between the cold, water and wood, that the tree has to protect its sap. The tree needs water, yet water makes it vulnerable to the cold. Water at its core, its spine: delicate, fragile and vulnerable.

Penpont, Dumfriesshire, 28th December 1995

The weather here became intensely cold, far colder than I have ever known it before. I know this because of the ease with which I was able to freeze icicles. I worked at Glenmarlin Falls where I last worked similarly with icicles in 1987, but even then I remember it being much more difficult. Now I was able to break and freeze icicles with a speed that I have never been able to do before in Britain. All that I did in Alaska flowed into the work that I made here. What I had learnt about the cold and the tree continued. I reconstructed icicles using the tree. After Alaska the tree became much, much more than a support for the icicles. The icicle is an expression of something of the core of the tree, its spine, and the works that I have made are as if for a moment that spine has shifted and stepped outside the tree, almost as an apparition.

Poppy petal wrapped hazel branch
held with water
raining

STONEWOOD, DUMFRIESSHIRE
1 SEPTEMBER 1992

Branch and hole

RUNNYMEDE, CALIFORNIA
30 OCTOBER 1992

Wood through wall
field gate reworked

Stonework by Joe Smith

BUCKHORN, NEW YORK
NOVEMBER 1993

Mulga branches
laid in two directions
changing with the light
as the sun moved round

MOUNT VICTOR STATION, SOUTH AUSTRALIA
28 JULY 1991

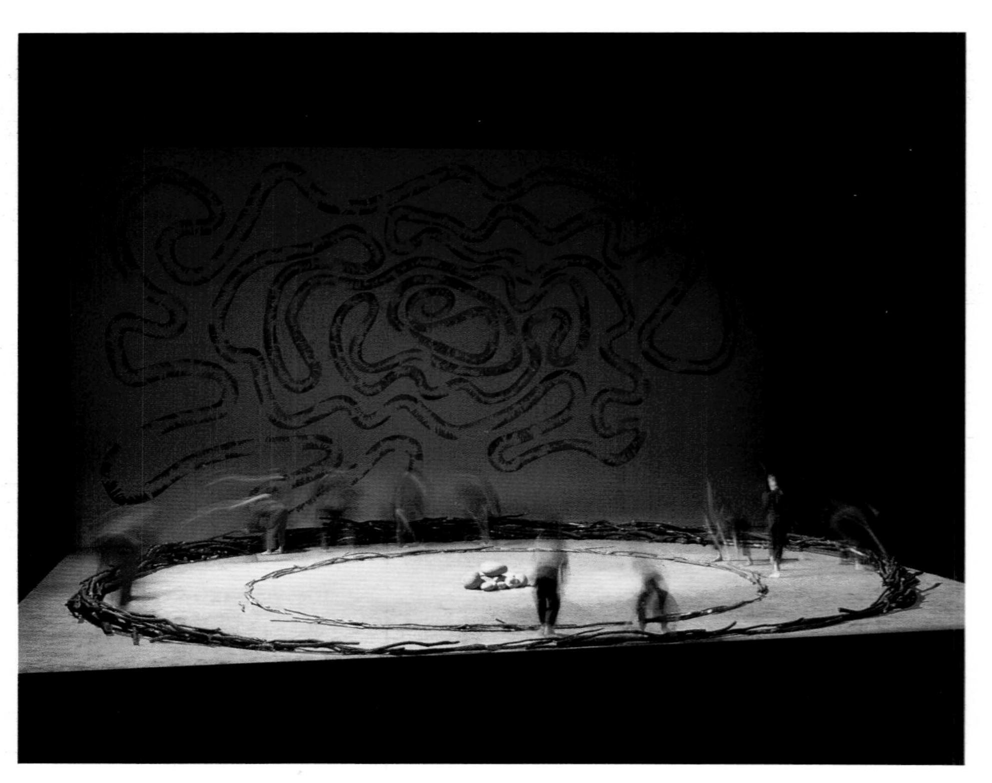

'Végétal'
stick house

LA ROCHELLE, FRANCE
NOVEMBER 1995

Ballet Atlantique
stick throw

PENPONT, SCOTLAND
JULY 1995

Sticks
joined by freezing one end to another
used the same stump for the first two works
many collapses
easier to work before sunrise and after sunset
twelve sticks in the first work

twenty-three in the second
both taken down intact
incorporated into a third work (*overleaf*)
one hundred and fourteen sticks

ANCHORAGE, ALASKA
26, 27-28, 29 NOVEMBER 1995

Ice houses
stacked sticks
poured icicles

Assisted by Cécile Panzieri and Julie Cruz

ANCHORAGE, ALASKA
8, 9 DECEMBER 1995

The coldest I have ever known in Britain
able to work and freeze icicles as never before
started early
worked all day
reconstructed icicles
around a tree
finished late afternoon
catching the sunlight

GLEN MARLIN FALLS, DUMFRIESSHIRE
28 DECEMBER 1995

LEAF

Penpont, Dumfriesshire, November 1989

At first I used only windfallen leaves, relying on the wind and frost to provide materials. I have been in woods after a hard frost in autumn with trees noisily shedding leaves as the sun rose. There is a strangeness in leaves dropping to the ground on a calm day, unlike the wind that can rip the leaves away from the tree. The wind that brings down materials to work with causes problems in the making. I have worked in dips, made windbreaks, only to have work that was nearly finished blown away by a sudden gust from an unexpected direction. It took several years to realise the potential of bracken slivers and thorns as a way of securing leaf to earth. The thorn in turn became the means to lift the leaves off the ground.

Gradually I began to use leaves from the tree as well as the wood floor. Not out of impatience but because the leaf on the tree has a different quality. To understand leaves, I need to work the dry brittle windblows, the cold wet frost-fallen, and the fresh green growing. I am careful about what I take. A few leaves from each tree.

I have learnt that thick leaves are found on the outer branches where the sun reaches and growth is strong. Towards the trunk the leaves are spring-like and delicate. The largest leaves are found on smaller trees, on trees that have been stooled, with roots that have more vigour than branches. The sycamore has taught me most, the biggest lesson being that so much can be found in something common and ordinary. Its leaf can turn all colours; its stalks can go bright red and within its leaf structure I realised my first leaf construction. By using its architecture I can make forms that grow from the leaf. The sycamore makes boxes, the chestnut makes spirals and horns, an exploration of structure and growth.

Sweet chestnut leaves
in a black basalt libation bowl
(about 625BC)

EGYPTIAN SCULPTURE GALLERY, BRITISH MUSEUM
OCTOBER 1994

Hickory leaves
walled round
river dry
raining and flowing two days later

LAUMEIER SCULPTURE PARK, MISSOURI
JUNE 1994

Line to follow colours in maple leaves
pinned to tree with thorns
first snow
fall meeting winter

DUNBAR, PENNSYLVANIA
18 & 19 OCTOBER 1992

Overleaf
Elm leaves
stitched together with stalks
making a line
from tree to river to tree

SCAUR WATER, DUMFRIESSHIRE
12 OCTOBER 1994

Maple leaves
pinned with thorns
between two trunks of a tree

PLANO, ILLINOIS
24 OCTOBER 1992

Horse chestnut leaf stalks
thorns

TOCHIGI PREFECTURAL MUSEUM OF FINE ARTS, JAPAN
OCTOBER 1993

Rushes
thorns

HAINES GALLERY, SAN FRANCISCO
JUNE 1992

Snowball

Japanese ferns (top)

Scottish ferns and bracken (below)

Horse chestnut leaf stalks

thorns

Photograph (*above*) Norihiro Ueno

SETAGAYA ART MUSEUM, JAPAN

FEBRUARY 1994

TREE

Capenoch, Dumfriesshire, January 1996

The long branch that has grown horizontal to the ground has taught me that the tree is the land. The branch is like the landscape. It is the earth, it is stone – these things flow through the tree as they do through the field and the mountains. Many of the works have dealt with the space occupied by the branch: there is something protective in the first gesture it makes as it breaks from the main trunk, growing upwards, then down in an arch, and I have made works that have sat in that space, not touched by the branch, but very much about it. I have worked best when there have been no leaves on the tree and the tree is reduced to its structure. The most difficult time is in mid summer when the tree is in full green leaf. I hope one day to make work that touches and understands better that time in the tree's life. This year I realised why I was having such difficulty. I kept trying to make works on the branch similar to ones made in other seasons, but the light killed off their form and colour and I realised that it wasn't the branch that was important in summer, but the shadow cast by the tree and that I should look to the ground and work with that shadow. Unfortunately, last summer was the driest for a long time and the ground was hard and I only managed a few scrapings in the hard dry earth. This is a direction that I will pursue in future.

The branch faces east and is well lit by the rising sun, which is often when a work is at its best; inevitably this is usually when I am just beginning. So with works such as the dandelion lines there was a fine edge between succeeding and failing.

Capenoch, February 1996

Decided to make a work on the tree with ice. Going dark – it didn't feel too cold and I was anxious about whether or not the ice would freeze. My friend Wallace Gibson told me that the weather forecast was only for -1°C, which was not really cold enough for doing such a work, so I gave up and went back home. But at about nine o'clock it seemed to be freezing so I returned to the tree. My intention was to build a pointed arch incorporating the branch where it dips down closest to the ground in a 'v'. It was such a beautiful night – clear, full moon – easy to see. Getting the ice from the pond was quite strange because the surrounding trees meant that it was often in shadow. It was very dark, but I managed to wade out quite far to get the necessary pieces. I used much more than there is in the arch because a lot broke up into pieces too small to be used. There were several collapses, two major ones in the lower part. I was able to build the arch in two sections, the top half and the bottom half separately, which enabled me to work more quickly. Even so, I worked from nine o'clock until four-thirty in the morning before it was finished. I had worked with the moon as it rose behind me and until it was setting in front of me – it was beautiful to see the light shining on the ice. It was an extraordinary night, so intense, – the light, the ice, the cold, the tree. The arch turned out much taller than I had anticipated, over seven feet. I had to stretch to reach the last few pieces – that was the most difficult bit. If you let a piece slip, the ensuing shock can break the arch. There's a real feeling of anticipation and tension as the two sides come together. I used very little support – only a few small branches to act as a prop – relying on the freezing of ice to ice to branch.

I photographed the work by moonlight, but the exposures take so long that by the time I'd done only a few, the moon had slipped down. I came home, slept for two hours, set my alarm clock and went back at dawn. A surprise to see it in the daylight! This is a very good piece, I am very happy with it and although I am tired, the tiredness is more than compensated by the joy of having achieved this work. Works like this are what I live for.

Branches
stacked
to form an opening
into the tree
at its trunk
from where it grew
snow a few days later

for Julia

16 JANUARY 1996

Overleaf
Snow line
melting during the day
freezing at night

2 FEBRUARY 1996

Began on Saturday evening
worked through the night
took ice from nearby pond
several collapses
finished early morning
full moon

3, 4 FEBRUARY 1996

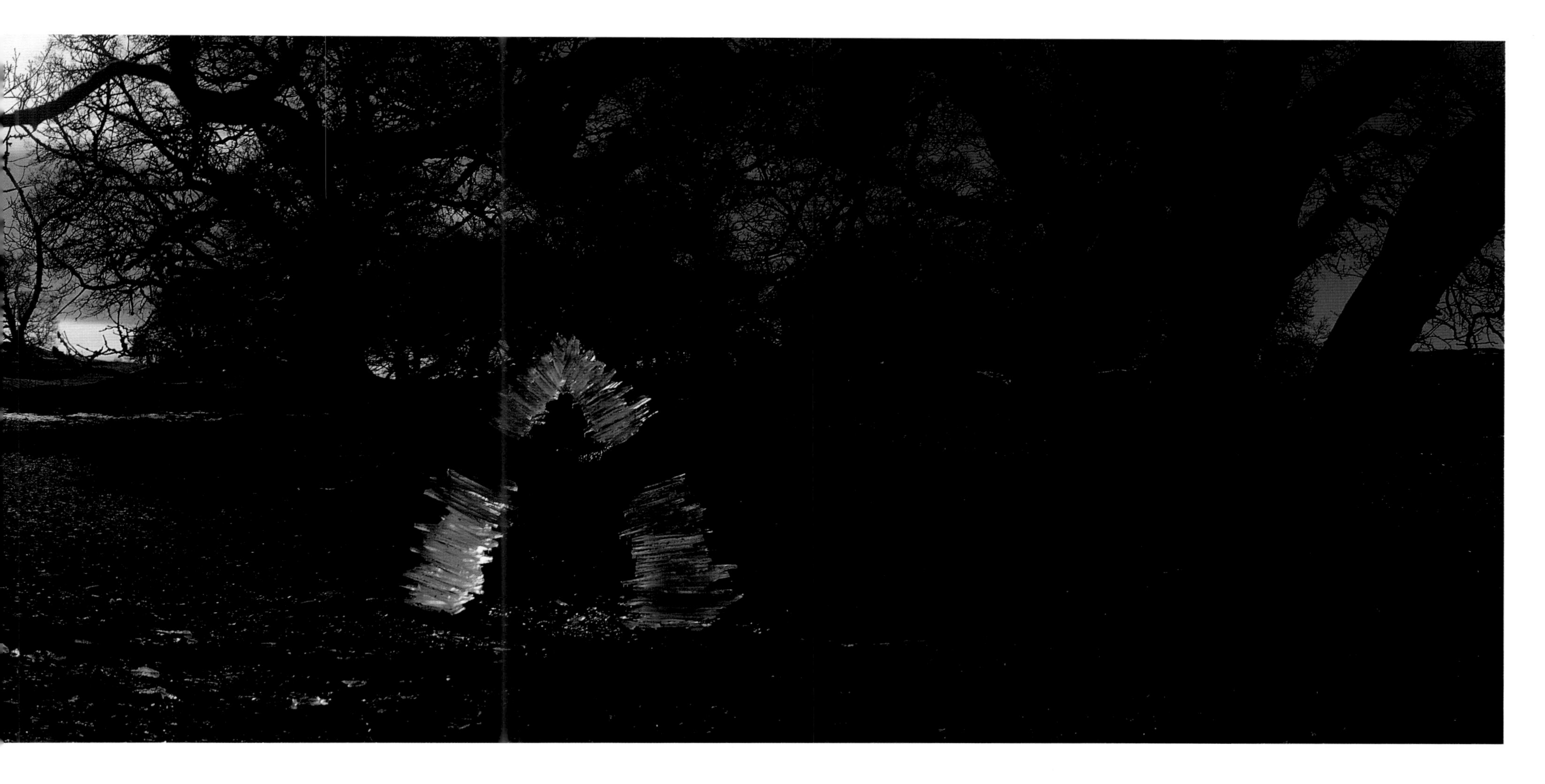

Filled in under the branch arch
with snow
made a hole
that changed with the light

8 FEBRUARY 1996

Smeared mud around the lip of the hole the next day (*overleaf*)
before reworking the snow
below the branch
forming an edge, with mud

9 FEBRUARY 1996

Rosebay willowherb stalks
thorns

25 FEBRUARY 1995

Five stone arches
raining heavily
muddy
windy
one collapse

MARCH 1994

Hollow snowball
wet, rapidly thawing snow

7 MARCH 1995

Strips of well-rotted oak split into short lengths
pinned with thorns to make lines

APRIL 1995

Waited for the dandelions to open
collected quickly
made a line along the branch
before they wilted
sunny

MAY 1994

Two holes
one with leaves
one with earth

JUNE & SEPTEMBER 1995

Oak leaf lines
stitched together with grass stalks
intermittently calm and windy

JULY 1994

Stone houses
balanced river stones

AUGUST 1995

Collected the brightest orange
windfallen leaves
line stitched together with stalks
pinned to the branch with thorns

OCTOBER 1994

Stacked sticks
to make
cone, ball, column
around, under, through
the branch

NOVEMBER, DECEMBER 1994

Overleaf
Seven snowballs
melting
returned the following day

4, 5 JANUARY 1995

Oak tree snow cairn
capped with mud
mud covered

snow covered
each another layer
of same work

10 FEBRUARY 1996